COUNTING ON A
SMALL PLANET

COUNTING ON A SMALL PLANET

ACTIVITIES FOR ENVIRONMENTAL MATHEMATICS

ANN & JOHNNY BAKER

Heinemann
Portsmouth, NH

Heinemann Educational Books, Inc.
361 Hanover Street
Portsmouth, NH 03801-3959
Offices and agents throughout the world

Library of Congress Cataloging-in-Publication Data
Baker, Ann.
  Counting on a small planet: activities for environmental
  mathematics

  p.   cm.

  ISBN 0-435-08327-9

  1. Environmental protection — Mathematics — Case Studies
  2. Ecology — Mathematics — Case Studies.   3. Human ecology —
  Mathematics — Case Studies.   I. Baker, Johnny. II. Title.
  TD170.2.B35 1991
  628.01'51-dc20
  91-30571   CIP

Published simultaneously in 1991
in the United States by Heinemann
and in Australia by
Eleanor Curtain Publishing
2 Hazeldon Place
South Yarra VIC, Australia 3141

Production by Sylvana Scannapiego,
Island Graphics
Designed by Sarn Potter
Cover design by David Constable
Cover photograph by Michael Curtain
Text photographs (except where otherwise
credited) by Michael Curtain
Illustrations by Kon Wong
Typeset by Optima Typesetters
Printed in Australia by Impact Printing

CONTENTS

INTRODUCTION

◆

It has always concerned us that maths is something that happens mainly inside the classroom with children seated at desks or around some objects on the floor. Why, we asked ourselves, can't maths be more active or more like an adventure? Why does it need to be sombre and desk-bound? At least once in a while, could it not be presented or practised in a less serious way and in physical contexts outside the classroom? We were not looking for any decrease in mathematical rigour but rather an increase in engagement, enjoyment and relevance in the maths that children do.

Our first thoughts then were:

'Where else can children do maths?'

'Why would they want to do maths elsewhere?'

It was while we were pondering these questions that we visited a classroom where the children were researching what happens to their rubbish after the refuse collectors take it away. Some data — facts and figures — had been collected but it was obvious that these Year 3 children had no idea what a family's 3 tonnes (3.3 tons) of rubbish a year might look like. Later that day the class teacher told me that she was anxious to get maths into her rubbish theme and showed me a few workcards she had already prepared, for example:

'If your family uses two cartons of milk a day, how many cartons is that in a fortnight?'

The children, as you might imagine, were not very enthusiastic about the problems. That evening at home, we began to brainstorm the possibilities of a 'rubbish' theme and found that it could be very rich.

From this beginning, we found that it was not hard to imagine how much maths would be involved in understanding and protecting the environment. And, of course, this type of maths would largely happen outside the classroom, be relevant to all of the children, be purposeful

1

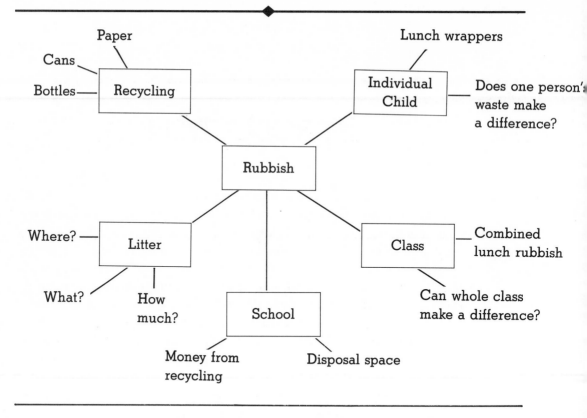

— especially if they reported their findings — and become very engaging to the children because the outcome would have real meaning and implications for them.

From our desire to find ways of moving maths out of the classroom into the real world came the notion of 'green maths', where the children would go further than using the environment as a source of mathematical activity; go beyond estimating the height of trees or exploring the symmetry of leaves. In 'green maths' activities, the children would use mathematics to understand the many environmental issues that face today's society.

MATHS AND THE ENVIRONMENT

These days adults and children too hear a great deal about the need to protect our environment. Most people give a lot of thought to the problem and yet too few follow through with any sort of action. Talking to friends and to children in classrooms about what they do to help all too often promotes the following type of response:

> 'What can I do that will make a difference? If we don't all do it there is no point.'

This leads to complacency . . . or is it laziness?

As we researched areas of interest for this book we again talked to people about the environment and discovered that perhaps one reason why many people are doing so little is that they just don't understand it. They might appreciate it and find it attractive but they have very little underlying understanding of how it is maintained and how it functions.

Discussion and reading can help people to broaden their understandings, but actually exploring the environment and using maths to explain or illustrate some of what's said or recorded could be much more powerful in changing people's awareness of themselves in relation to their environment. To put this claim into a real context and show its power we'll tell the story of one inner city school which had a problem with litter.

To reach the Year 2 and 3 classrooms the children had to walk along a narrow pathway that flanked a main road, from which it was separated by a chain wire fence. Each morning the children scrambled over mounds of litter. This was rather unpleasant for them and they began to complain. We asked them what they thought they could do to tackle this problem. The obvious answer was that the caretaker should clear up the mess before the children arrived at school rather than at about 10 o'clock each morning. This was not well received by the caretaker who listed all the early morning jobs that he already had. Think again. As the children thought a question arose:

'Whose litter is it?'

The children were confident that it was passers-by who were causing the problem, so they decided to collect and categorise the litter. Over many days the children did just that. They graphed their information and discovered some clear trends. A lot of the litter was made up of lolly wrappings, drink cartons, cigarette wrappings and so on, things that were not allowed in school. So where did it all come from? Most of the remainder was school litter — snack wrappings and scrap paper.

The children reflected on their findings and explained which litter there was most of but soon realised that they had not answered their initial question:

'Whose litter is it?'

and new questions were arising:

'How/why/when does it get there?'

The Year 2 and 3 children combined to decide what needed to be done to answer these questions. The following actions were decided on:

1   Get to school early and clear up the litter.
2   See how much litter collects between 8.30 and 9.00, the busy time when children come to school.

3   Clear up and repeat this process during the day at one-hour intervals and clear up at the end of the day when everyone has gone.

4   Spy on children at breaks to see what they do with their rubbish.

During this phase the children were collecting, sorting and recording their findings. Some interesting recommendations emerged from this, including changes to the siting of rubbish bins around the school. In fact since the children had noticed that wind direction and force affected the litter build-up they suggested moveable bins rather than ones fixed to the walls.

The children became annoyed because the non-school litter was coming from the corner shop. Investigation showed that there was no litter bin at all outside the shop. The children wrote explaining their concerns to the shopkeeper. The shopkeeper ignored this complaint. The children had almost eliminated the school litter problem and now felt quite angry about the non-school litter. For a whole week they collected and graphed this litter build-up. They also decided to weigh it and measure it in terms of how many black plastic bagfuls. A report with hard data was drawn up and presented to the shopkeeper with a threat to return all the rubbish and dump it on his doorstep. It worked: the shopkeeper provided litter bins outside. The whole school's awareness of litter was raised by the Year 2/3 project and presentations and the litter problem was almost solved. As for the mathematics, to solve the problem the children had:

- collected, sorted, and tallied the rubbish collected,
- represented their results as a block graph, using Lego to help them count each day's rubbish and transferring the results at the end of the day to graph paper,
- used clocks and timers to indicate when the next collection was due, and written the times of collection next to their block graph,
- made a plan of the school grounds, for use in recording data collected and for locating litter black spots,
- discovered the need for NSEW and intermediate compass readings in order to record the wind's direction, as well as developing an informal scale for measuring the wind's strength,
- presented their findings as a combined report, using mathematics to explain the issues and persuade their audience.

The teacher and the children in the class were satisfied with their efforts, and felt they had learned a lot. The litter problem was brought under control, even though, as they say, you can lead a horse to water but you can't make it drink. Some passers-by were still observed dropping litter.

This example shows how even young children can use mathematical techniques of data collection and interpretation to:

- find out why a problem exists,

- work out ways of solving it,
- make sense of some aspects of the environment as it relates to them,
- develop positive attitudes to preserving their environment,
- improve the situation.

These points also form the focus for this book. It has long been our view that a major goal of primary education is to prepare children for responsible citizenship, to give them the ability to understand social issues and to make personal, well-informed decisions. We also feel that maths has a vital role to play in developing this ability, and decided, therefore, to develop a series of activities in which environmental issues were combined with maths education. (See note, page 11, for age-ranges for these activities.)

WHOLE MATHS, WHOLE LANGUAGE, WHOLE LEARNING

Our earlier books focused on problem solving, process maths and the development of mathematical strategies. As we began planning to tackle the theme of mathematics in the environment, the ideas for activities proved easy to come by. Indeed, so excited were we about the possibilities that we were soon designing plans for action:

'The children could do . . .'

'They could set up an experiment to . . .'

'It could be measured by . . .'

We soon realised that this approach was not going to result in any of the child-centred, process-based activities that we still believe are at the basis of successful maths learning. Alarmed at how easily we had let the content of each topic (and our ideas of what would be the best way to mathematise it) dominate our thinking, we felt it important to re-establish the educational goals of the material: learning about using maths in the environment, yes, but equally important, developing as a young mathematician, i.e. learning about maths as well! The outcome, we hope, is that this book is more than just a collection of useful, relevant and enjoyable activities. It also suggests starting points from which the children in your class will generate their own questions, investigations and purposes.

The activities in this book encourage natural learning, as they are based on the conditions that are present when children learn naturally. In this way, the book takes a 'whole' approach to learning. Whole language is a term familiar to most of us and brings to mind images of classrooms where:

- learning in the language arts is allowed to happen naturally because those conditions that facilitated the learning of language in young children are recreated in the classroom,

- language is kept whole — learning is not broken down into isolated skills and a sequence of steps,
- learners are empowered because they are encouraged to make decisions and take on much of the responsibility for their own learning,
- language is used in meaningful process-oriented not product-oriented situations where what is being learned is seen to be purposeful,
- a wide range of contexts, approaches and resources facilitate language growth,
- talking, listening, reading and writing are integrated into one language strand.

An aspect of 'whole language' that is not often seen is that of whole language spanning all curriculum areas. In most classrooms after the language arts lesson whole language switches off and gives way to maths, science, social science and so on. For some time now we have been grappling with the idea of 'whole' maths. Whole maths is not just whole language about maths. Simply reading, writing, talking and listening in the maths lesson will not result in a whole maths lesson. If we take each of the above points in turn, though, and apply them to the maths lesson, we can create a classroom where:

- learning about maths is allowed to happen naturally — natural learning conditions such as immersion, demonstration and expectation are created in the classroom,
- maths is kept whole, i.e. concepts are not broken down into isolated skills to be practised in sequence,
- learners are empowered because they are encouraged to ask questions, to make decisions about the approach they will take to a problem and to take responsibility for their own learning,
- maths is used in meaningful situations and what is being learned or practised has immediate use and relevance,
- a wide range of contexts, approaches and resources foster mathematical growth.

What results then is a situation where whole language and whole maths can be integrated, not just in the maths lesson but across other curriculum areas too. The very theme of this book — our environment — might be considered to belong more to the science or social science lesson but we believe that what is presented here is actually a way of getting started on 'whole maths learning'. It would be quite feasible to introduce one of the topics presented and for it to encapsulate a curriculum for a full day, or even a week, without imposing on it any of the restrictions that often accompany a thematic approach. The topics can take on a life and momentum of their own, making possible 'natural learning'.

THE PROCESS APPROACH

Whether they have a problem to solve or an investigation to carry out, children find themselves embarking on a journey which is guided by a 'process'. The process will be their overall strategy which they use to decide:

- how to get started,
- what actions to take,
- how to gather the information they need,
- how to report results, whether intermediate, conclusive or even inconclusive, to others.

With help from their teacher, they may also find time to reflect on what they have achieved, learned and now need to learn. As we observed children working on the activities that form the case studies section of this book, we found that it was helpful to record what happened under the 'process' headings that we have found so useful in other contexts:

Experiencing the Problem

- Incubating
- Brainstorming
- Questioning
- Discussing

Mathematical Activity

- Calculating
- Formalising
- Researching
- Summarising
- Observing
- Constructing
- Diagramming
- Interpreting
- Generating data
- Drawing conclusions

Communicating Results

- Talking
- Listening
- Writing
- Reading
- Reporting
- Reacting
- Discussing

Reflecting

- Knowing what you know
- Knowing how you know it

These aspects of process are not always followed in sequence, but the four headings do provide a valuable framework for helping the children to undertake an investigation effectively.

CASE STUDIES

- ◆ All That Rubbish
- ◆ Be Quiet
- ◆ Water Use and Abuse
- ◆ Not a Drop to Spare
- ◆ When the Wind Changed
- ◆ Design a House

This section consists of environmental topics that we have trialled with children. In the accounts, we have emphasised how the children engaged with the topics, and how their questions became the focus for investigations. Because the children had posed the questions, they had a personal involvement in finding answers and took pride in communicating to the rest of the class the information they had uncovered. This method of working also relied on collaborative group work. In almost all cases, we were able to set up small interest groups of four or five children who, having expressed a desire to follow a particular line of investigation, worked harmoniously together. Occasionally it was necessary to allocate children to a particular group, but even that process was carried out with their consent, by negotiation, offering a choice, rather than by directing them.

To make it easy to see how the case studies unfolded, each has been presented as follows:

ENVIRONMENTAL POINTS

To help the children engage with a particular issue, we have listed a series of environmental points for general discussion. These can help break the ice and quickly establish the topic that follows.

A BRIEF ACTIVITY

Particular aspects of a general topic can often be explored by the children within the scope of an activity that lasts for no longer than a single maths lesson. If their interest in the issue is sufficiently aroused by this activity, they will be ready to progress further. If not, awareness of the issue will have been raised and might develop later, particularly if something happens (such as a sudden storm) to make the issue relevant to them.

POSSIBLE OUTCOMES

When we looked at the avenues that were explored by the children we worked with, we found each issue was taken beyond the initial activity. In all cases, we encouraged the children to pursue their own interests, providing them with opportunities to gain a broader perspective on the topic by listening to report-back sessions from other groups in the class. The general aim of this section is to describe what is possible, rather than prescribe what the children should be doing.

MATHEMATICAL CHECKLIST

To satisfy ourselves that each case study is viable from the mathematics point of view, we looked back over the work done by the children and analysed the maths that had been used. Doing this gave us the kind of evidence that many teachers need before devoting valuable class time to an extended project of this general kind.

FACT FILE

Children have a fascination with 'facts' and large numbers. The fact files were introduced after the children had done some investigation and were often used to compare local findings with more global information.

Note: These activities were trialled with children aged from 5–12 years. As we worked with children of different ages we found that what may appear on the surface to be inappropriate for children of a particular age-range was easily adapted to meet their needs. Most importantly, remedial children or those in need of extra support found the activities of great interest and responded very positively to the challenge of studying their environment through maths.

ALL THAT RUBBISH

ENVIRONMENTAL POINTS

1   We all generate and so throw away more rubbish than we need to.
2   Reducing the amount and type of packaging and waste products will help our environment because we will:

   - need to cut down fewer trees,
   - use less fossil fuel in the production of packaging,
   - reduce the amount of air, water and land pollution.

3   Individuals must take responsibility for their own actions and so persuade other individuals and producers to be more conscientious too.

WHAT'S IN A LUNCH BOX?

How much rubbish do you have left in your lunch box when you've eaten your lunch?

What kind of rubbish do you have left?

How much is that in a week/month/year?

Which rubbish don't you need to have?

How much less could you throw away?

How much is that in a week/month/year?

EXPERIENCE OF THE PROBLEM

A session such as 'What's in a Lunch Box?' provides the children with an initial experience of the problem, and they may want to pursue this topic further. The class we worked with wanted to expand the basic idea to include all forms of rubbish that they created. In particular they were interested to find out:

'How much rubbish does the class throw away in a week?

'What kind of rubbish do we throw away?'

'Could we cut down on the rubbish we make?'

'What is recycling?'

'How much of our rubbish could be made into something?'

'Could we go for a week without making any rubbish?'

This list was part of a longer set of questions that the children thought of as a result of the previous activity. Alternative ways to give children further experience of a 'rubbish' theme could include:

- having a litter drive or a 'Keep Our School Tidy' campaign,
- reading books with a 'rubbish' theme, such as *Dinosaurs and All That Rubbish,* or *Bottersnikes and Gumbles,*
- collecting and discussing newspaper articles on topics related to 'rubbish',
- visiting a local rubbish dump.

MATHEMATICAL ACTIVITY

When interest in the topic has been aroused, you will find that the children's ideas are so plentiful that three or four groups can be set to work on different aspects simultaneously. For each group it is important that they set a clear goal for their investigation, as this will enable them to quantify their aspect of the problem and to collect and analyse data that will help them reach a conclusion. The class that we worked with decided on the following areas for investigation (see table at top of next page), and separate groups pursued each area independently.

WASTE PAPER

The waste paper group asked each child in the class to put any pieces of paper that they had finished with into a central collection box. After two days, there was enough in the box for analysis to begin.

The children noticed that much of the paper used for notes or rough work was being used on one side only. They calculated that about one

Group	Topic
Waste Paper	We think our class wastes too much paper so we want to see how much we do waste and where we can avoid waste.
The Kind of Rubbish	We want to see what kind of rubbish we throw away each day and how much this is in a week or a month.
Rubbish for Recycling	We are going to sort the rubbish for recycling to see if we can sell any of it.
Rubbish Today	We are going to collect today's rubbish and then see if we can make less rubbish tomorrow.

third of the waste paper could be saved if the class got into the habit of writing on both sides of the paper used for rough work. On the other hand, they accepted that final work should be done on good paper, and that a certain amount of waste 'if you made a mistake' was allowable.

THE KIND OF RUBBISH

All the children in this group collected their own rubbish that day and found their own methods of recording their findings. These included:

paper	plastic	alumnium
scrap	popper	foil
scrap	nutella	
brown bag		
lolipop stick		

The components were weighed and the total calculated:

paper — 30 gms
plastic — 20 gms
aluminium — 5 gms
wood — 3 gms
food — 20 gms
_____
total — 78 gms

The group then went on to estimate a 'total for the week' from their results, bearing in mind that some variation was possible, for example:

'I only get a drink on a sports day.'

RUBBISH FOR RECYCLING

This group identified the types of rubbish as 'stuff' that:

- can be recycled,
- can be reused,
- cannot be reused.

Since the school did not then have a policy for sorting its rubbish, the children made predictions, based on the 'Rubbish Today' group's findings of how much paper, aluminium, glass, and plastic could be expected in a week. They then planned to contact relevant agencies to see what was the minimum for collection and whether there would be any profit in it for the school.

RUBBISH TODAY

This group produced the first surprise for the class when they presented their tallies of 'today's' rubbish. Everyone became committed to contributing less to the rubbish pile 'tomorrow'. The group then had the task of devising a way of showing the two collections in a way that would enable the rest of the class to see how successful they had been in reducing the waste. They experimented with a graph, but eventually settled on a visual comparison ('Look how much less ...') backed up with the weight of each day's waste. The class could see that, with a bit of effort, serious inroads into the rubbish problem could be made.

So interested was this group in their findings that they decided to continue their work. Drink cartons became their major focus, and the mathematics of their investigation was included in their report to the rest of the school (see below).

COMMUNICATING RESULTS

It was decided to present the class findings to the Junior section of the school on Friday afternoon. The classes were assembled and, with the awful truth glaring at the children in the form of three bags of rubbish, the related quantities and masses were read out:

> 'There's one whole sack of drink cartons here. There are two hundred and twenty-one in this sack. That means we each throw out about one and a half cartons each week. Two hundred empty cartons is 2 kilograms (4.4 lb). If we all brought drinks in plastic containers that we reused that would save this sack of rubbish.'

To make their points stick, the class announced that they intended to inform the rest of the school about how much rubbish was being made by 'publishing' their results on clipboards at each disposal site. They also planned to write to the local paper about what they had found out — although fear that this might reflect badly on the school eventually changed their mind on this.

REFLECTION

When asked:

> 'Were there any surprises?'

the children all stated that they had no idea how much rubbish they generated and all felt that, yes, they could make a difference. Fired by this they now had some ideas about what they wanted to find out next. These included:

> 'Let's see if we can make less rubbish next week.'

> 'Let's see if we can make half as much rubbish next week.'

> 'Let's estimate how much the whole school uses.'

> 'Let's report to the whole class and keep all the rubbish next week so that we can show them how awful we are.'

> 'Let's get our parents in to see that they can make a difference.'

> 'We'd like to know how many trees our class uses in a year.'

'We want to work out how many aluminium cans our class uses in a year, and then see how many times they can be recycled until there are none left. How many cans would that be altogether?'

MATHEMATICAL CHECKLIST

As this class became enthused by their rubbish topic we noticed that they:

- moved naturally into maths to help them understand it,
- recorded information that they had collected,
- used this mathematical information to demonstrate what the problem was and to persuade others that they should join in too,
- made sense of an environmental issue that related to them.

During the course of this topic you could expect the children to:

Mathematical Topic	Usage
Sort and classify	By material (plastic, glass, paper etc.)
Order and compare	By mass, volume and quantity
Add and multiply	To keep totals of collections
Use fractions	To compare categories of rubbish
Invent methods of recording	To represent data for daily and weekly collections
Estimate amounts	To make long-term predictions
Calculate mass/volume	Inventing methods to find the mass of small objects
Use a calculator	To check calculations, e.g. long-term predictions
Make ground plans	To position collection sites and to post findings
Use graphs	To summarise daily and weekly findings
Incorporate mathematics into written reports	To persuade and explain to others

A general goal for this kind of topic is for the children to discover for themselves that mathematics is a powerful tool in the preparation of a report or in presenting an argument for change.

FACT FILE

Each Australian throws away 700 kilograms (1500 lb) of rubbish a year. Without recycling, this amount could double in the next twenty years.

Each American throws away about 27 kilograms (60 lb) of plastic packaging every year.

There is an estimated $8 million dollars worth of reusable aluminium and glass thrown away in Australia each year.

Americans use 2.5 million bottles every hour.

Every day Americans use 35 million paper clips.

Household garbage includes 30 per cent paper, 12 per cent plastic, 9 per cent steel, 7 per cent glass, 1 per cent aluminium, and much of what remains is compostable.

Worn-out batteries (especially little watch batteries) are a major source of mercury pollution.

The manufacture of a battery uses 50 per cent more energy than the battery can store. Wherever possible it is less wasteful to use the mains supply.

Plastic waste, thrown into the sea, kills 30,000 seals each year.

Did you know . . .

- A process called neutralysis has been developed in which 1 tonne of clay, 1 tonne (1.1 tons) of garbage and 300 litres (66 gallons) of liquid are used to produce 1.3 tonnes (1.4 tons) of rock-hard building pellets.
- The energy taken to produce three hundred aluminium cans would run the average family car for a week.
- The reason why glass and aluminium companies pay for the return of their products is because it is cheaper to recycle than to produce from raw materials.
- The chlorofluorocarbons from old refrigerators can be reused.

BE QUIET

◆

ENVIRONMENTAL POINTS

1   Noise pollution can cause stress, lack of concentration, headaches and over extended periods of time, deafness.

2   Awareness of the effects of noise and which noises cause most stress will help us to:
- reduce the amount of unnecessary noise pollution,
- consider the effects of our noises, time, duration and so on, on others,
- understand why we are irritated and annoyed by noises.

3   As neighbours, workers, co-workers, friends and members of the broader community we can perhaps reduce the overall noise pollution and related stress, even at the level of:

- not suddenly shrieking or beeping a horn,
- not playing our music too loud for neighbours or at the beach,
- giving the mowing a miss at 6 o'clock in the morning.

This page may be photocopied free of copyright.

20

NOISE IN OUR CLASSROOM

Which noises in your classroom are most distracting?

How could you measure which is most distracting?

What equipment will you need?

Where/how/when will you carry out your test?

How will you present your findings for best effect?

EXPERIENCE OF THE PROBLEM

A session such as 'Be Quiet' may generate a great deal of discussion and contention about which noises are most distracting and will promote many new questions that the children will want to investigate. The Year 4 class we worked with listed their questions on the board as they went along. These included:

'Which noises are annoying or cause a problem? Why?'

'When are noises annoying? Why?'

'What effects do noises have?'

'Which noises are the loudest?'

'Is the longest noise the most annoying?'

'When is it quietest? Is this also the time when it is easiest to concentrate?'

Having listed these questions, interest was aroused and the children wanted to begin investigating immediately. You may prefer to introduce this noise pollution theme in a different way, for example by:

- sitting very still and quiet and listening to all the background noises,
- reading books with a related theme, e.g. *Mr Noisy, The Bop,* and *Bertie and the Bear,*
- recording then playing back tapes of noises such as traffic, extremely loud music, a jackhammer, construction site, baby screaming, vacuum cleaner etc.,
- trying silent reading to a range of background noises to see what effect they each have,
- visiting a construction site, traffic works, airport, or factory where noise pollution is a problem,
- creating a 'noisescape', using instruments and improvised instruments to tell a story without words, e.g. getting up in the morning and going to school.

MATHEMATICAL ACTIVITY

The class that we worked with decided that their first task was to find out:

'How can we measure noise?'

The suggestions that they investigated in small groups included:

'We could measure how far away the noises can be heard, then we'd know which was the loudest.'

'We could make all the noises at once and see which noise we can hear above all the rest.'

'We could measure the vibration, make the noise down a tube and measure its movements.'

'We could do a survey, asking people which noises annoy them most.'

'We could listen to noises and rate them on a scale of one to ten, softest to loudest.'

'We could record them all through a microphone then play them back to see which makes more red lights on the "noisemeter".'

'We could draw how high/long the noise is.'

It took a while for the children to draw up clear plans and to collect the equipment required but then they were free to conduct their investigations. We knew that problems would arise with some of the measuring techniques but wanted the children to discover this for themselves. The first problem arose when the group measuring distance discovered that they could not get far enough away from the point of noises so had to find a different starting place. Paces were their chosen measurement and of course the three children in this group soon discovered that their paces were all different. A trundle wheel was fetched but even this presented problems. Each time they measured, they began again from base, which took quite a while and was soon seen to be inefficient. Markers at 10-metre (11-yard) distances were used and labelled. This helped and information was then able to be collected quickly.

The vibrations group discovered that the tube did not vibrate at all unless it was used as part of the noise-making process so they had to think again. Making several noises all at once also proved problematical since no one was clearly dominant all the time.

COMMUNICATING RESULTS

At the report-back session the children were invited to discuss:

- what they had tried,
- what happened,
- any surprises that they had had,
- how they would now vary their investigations.

As each group completed their report, the other children were invited to comment and question. The questions asked were very perceptive:

'You stood down at the oval while the noises were made at the top of the hill. Would it be the same if the noises were coming up the hill or across the top or on the flat?'

'As you drew lines to match the length of the sounds how do you know you always drew at the same speed?'

'My dad says the lights flash in the tape deck because of the bass not the noise, what do you think?'

'We made a telephone with two cans and some string. Could you test vibrations like that?'

BACK TO THE DRAWING BOARD

Undaunted by the lack of clear answers the children were keen to keep going and had all decided that they needed to be more accurate with their measurements and measuring techniques. The group that were drawing lines to match the length of sounds decided to add measurement of tone too. Their idea was to mark 1 centimetre (.4 in) spaces on a strip of paper and mark it with staves like music paper. A metronome was collected and a notch on the paper marked to match each swing as the noises were made. They measured nails scratching on the board, chairs scraping, coughing, paper rustling and other common classroom noises. They went on to invite helpers to record simultaneously so that they could 'average' the results and get a more accurate measurement.

The two boys doing the survey had interviewed people, asking what their most annoying noises were. They had collected a long list of noises, some quite unusual, and realised that they had no clear outcome so decided to ask people of all types to rate the ten most common (i.e. occurring most often) noises giving each a value from one to ten, ten being the most annoying.

REFLECTION

Some clear results were now being presented to the class. The children commented that they had seen the point of having lots of different ways of measuring and thought it was interesting that there were so many different ways of measuring noise. As they shared their results, though, new questions were generated:

'Why are some noises more annoying than others?'

'Is the loudest noise also the most annoying?'

'Is the longest noise the most annoying?'

COMMUNICATING RESULTS

The class went on to investigate their new questions and also researched sound, how it travels and how scientists measure it. A classroom poster for everyone's attention was made, listing classroom noises in order, from the most to the least annoying. Suggestions were also made and published in all classrooms highlighting which sounds travelled most between classrooms and which could be heard most clearly from outside the classrooms.

REFLECTION

When asked:

'What surprises did you have?'

'What interested you most?'

'What do you know now that you didn't know before?'

the children all agreed that they really hadn't realised how annoying noise could be, or which noises were most annoying and why. They had discovered aspects of pitch, tone, frequency and intensity although they didn't use that language. Of course, they didn't want to stop there — new questions had arisen:

'How can we cut down on echo?'

'How can we eliminate or cut down on some noises?'

'What is a comfortable noise level?'

'How often and for how long should we have quiet times during the school day?'

MATHEMATICAL CHECKLIST

As this class worked on the 'Be Quiet' topic we noticed that they:

- confidently invented measuring techniques,
- used trial and error to improve their methods,
- saw the need for more formalised and standardised measuring techniques,
- invented ways to record and present the information collected,
- presented their findings to other classes to persuade them to reduce noise pollution.
- moved naturally from the real world phase of the problem into mathematics and then used their results in the solution of the 'noise' problem.

During the course of this topic you could expect the children to:

Mathematical Topic	Usage
Invent measuring techniques for:	
distance	How far away can noises be heard (paces, metres)?
loudness	Pitch, tone, intensity
passage of time	How long do the noises last?
a scale of intensity	1 - 10 from softest to loudest, informally using flashing lights on the tape recorder
Design questionnaires and surveys as a measuring tool	Collecting data, representing data, interpreting data
Invent methods of recording	Lists, tables, scales
Average their results	To get a clear picture
Research formal methods of measuring 'sound'	Decibels

FACT FILE

People who work in noisy environments like airports or with noisy equipment have to wear earmuffs or they might experience ear damage and suffer hearing loss.

In North Carolina it is against the law to sing out of tune, and in Little Rock, Arkansas, it is illegal for dogs to bark after 6 p.m.

There is a machine that can measure noise — the unit of measurement being the decibel. The more decibels recorded, the louder the noise. Decibels measure the intensity of the noise. The following table shows the intensity of some noises:

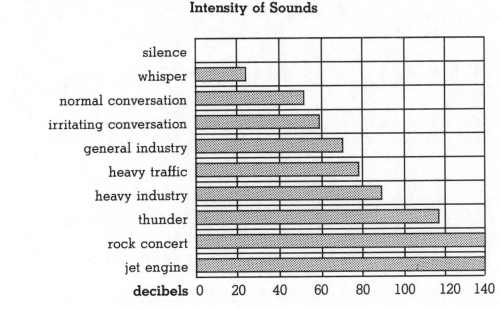

Intensity of Sounds

Did you know . . .

* Conversation at 50 decibels is comfortable, whereas conversation at 60 decibels is irritating.
* Safety standards for noise are set at 90 decibels for eight hours.
* Ninety decibels for long periods requires the use of earmuffs.
* Two or three minutes of a jet engine at 140 decibels can cause hearing loss.
* Loud noise or use of a Walkman at 110 to 120 decibels over long periods can result in deafness.

WATER USE AND ABUSE

◆

ENVIRONMENTAL POINTS

1 In many areas tap water has been found to be below standards of purity that are considered safe; this is largely due to overuse of limited resources. Because of this overuse, standards of water purification are decreasing and excessive amounts of chemicals are needed to purify water.

2 We all use far more water than we need to and this puts great pressure on water treatment plants.

Note: This activity looks only at the day-to-day use of water in our homes. The next case study, 'Not a Drop to Spare', looks at the use of water in the broader context of market gardening and farming.

WASH YOUR HANDS

How much water do you think it takes to wash your hands?

How will you measure the water used?

Will it be the same every time?

Will it be the same for everyone?

What equipment will you need to help you find out?

How could you use less water?

How little water could you use?

Design a poster to encourage people to use less water to wash their hands.

What will you tell your family?

EXPERIENCE OF THE PROBLEM

A session such as 'Wash Your Hands' may have aroused interest in the topic of water conservation. If it has, the children may want to turn this into a project and will, like the Year 5 children we worked with, look for other aspects of water wastage to explore. Questions raised by the class we worked with included:

'How much water is wasted when we leave a tap to drip?'

'How much water does a family use in a day?'

'How much is wasted?'

Alternative ways to introduce a water use and abuse theme could include:

- reading *Mr Archimedes' Bath* where there is water displacement to be considered and broader issues that could be introduced such as:
  . How full should we have the bath?
  . Is it more economical to have a bath or a shower?
  . How much water does it take to bath/shower?
  . What if there are two people to get clean?
  . What can we do with the water afterwards?
- reading *Tiddalick Frog Caused a Flood* which raises questions about how much we drink and how much we need to drink,
- holding a discussion with the children about how they use water at home. This might generate other questions that the children can investigate, for example:
  . How can we improve our water usage so that we use less water?
  . What are the big water users/wasters?
- visiting a reservoir or water treatment plant to emphasise just how much water is required to support our community.

MATHEMATICAL ACTIVITY

When the children's interest is high, they will want to choose an aspect of water use or abuse to investigate. Since water is such a valuable commodity, make sure that the children are clear about what their investigations are, how to carry them out, and what the point is. This does not mean tell them what to do, but they may need reminding to make clear plans that will help them collect, quantify and record information with the least possible water wastage.

The class we worked with decided on the following areas for investigation:

Group	Topic
Usage	How much water do we each use in one day/week/month/year?
Wastage	Which water usage wastes most water?
Washing Up	Is it better to wash up as you go along or wait till you have a lot?
Hosing	How much water does it take to hose the playground/path/car?

USAGE

This group began with a list of personal water use and made a matrix using tallying to show each day's water usage.

	Monday	Tuesday
Washing hands	⊞⊞ ‖	⊞⊞ ‖‖
cleaning teeth	‖	‖
Flushing toilet	⊞⊞ ‖	⊞⊞
Drinks	⊞⊞ ‖	⊞⊞
Shower	‖	
Bath		‖
Painting		‖
Water bombs		

They found out by investigation and research how much water each operation used and were then able to make daily records of each day's personal water usage.

WASTAGE

This group listed daily jobs where the tap was left running for a while, e.g. brushing teeth, washing hands and face, rinsing dishes. Their method was to 'measure' the amount of water that flows in one minute when the tap is on slow, medium or fast. From this they worked out how long each job took, estimated the possible amounts of water flowing and then compared that with the amount of water used in a bowl or with the plug in. Their findings were recorded as follows:

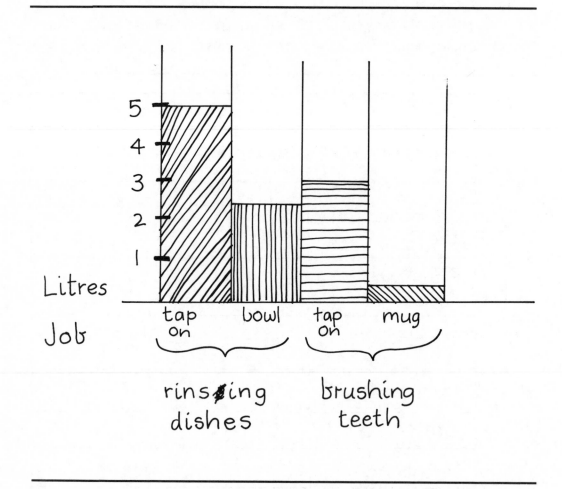

but the children felt that the impact was lost so experimented with other methods, for example:

Job	used	wasted	could use only	% water wasted
brushing teeth	1 litre	900 mls	100 mls	90%
washing up	2½ litres	1250 mls	1250 mls	50%

Over all we wasted about 75% less water

WASHING UP

This group took over the staff washing up for a while. Having interviewed the staff they discovered that the staff members washed their own cups at break times. They also discovered how many items each teacher used a day. From this they worked out an average amount of water used to wash one mug, one knife, fork and plate. A daily quantity was worked out and compared with one washing-up session at each break, or a once a day washing up, allowing for one change of water and for soaking water.

HOSING

Again this group's first job was to work out how much water was used in a given space of time if it was flowing gently, flowing moderately, or flowing with force. Timing the various jobs and noting the speed of flow was used as the basis for averaging and estimating results for each job.

COMMUNICATING RESULTS

The results of each group's work showed how wasteful and unthinking we all are in terms of water usage. The children thought that these findings were so important they wanted to inform the whole school and their families. Suggestions included:

- a roster in the staffroom designating washing-up duties,
- posters giving water-saving tips,
- water usage/waste fact books to be displayed in the school library,
- letters sent home with facts and figures to persuade families to conserve water.

REFLECTION

When asked to comment on their findings, the children particularly focused on their previous understanding of quantity, e.g:

'I didn't realise that water flowed as fast as it did. What I mean is when the tap's running a litre (1.76 pints) takes no time at all to flow.'

'I don't really think I even knew what a litre was like, so we spilt a litre on the path to see how far it would go. It was amazing, it went a long long way.'

'We thought that maybe it would take 500 millilitres (18 fl oz) to wash a mug. The mugs hold about 120 millilitres (6 fl oz) but to do a thorough wash and rinse takes more than a litre.'

'We experimented, with a little water in the mug, tap off, scrubbing the inside and outside then rinsing and it still took about 500 millilitres.'

'We also found that you use more washing-up liquid washing cup by cup.'

'We were surprised by how much water we waste and think everyone should just try these experiments.'

MATHEMATICAL CHECKLIST

As this class worked on this 'Water Use and Abuse' project we noticed that they:

- were prepared to take a risk, as they tried to establish reasonable methods of measuring water flow,

- used trial and error to improve their methods,

- improved their conservation of quantity,

- chose appropriate methods of recording information to emphasise the points they were trying to make,

- requested help in expressing findings as percentages,

- became proficient users of stopwatches,

- related their findings to the real world, proposing and supporting water-saving strategies.

During the course of this topic, you could expect the children to:

Mathematical Topic	Usage
Use a matrix	To represent water usage
Use formal and informal measurements, litre, bucketful, etc.	As they work out how much water is used
Combine timing with measurement to predict water usage	To find out how much water is used without actually wasting too much water
Find ways of representing their information	As a graph or table
Use percentages	To emphasise an argument
Use computation	To find daily quantities
Estimate	Speed of flow
Average	Quantities with different rates of flow

FACT FILE

The body needs 1.5 litres (2.6 pints) of water a day.

The average household in the Sydney district (New South Wales, Australia) uses about 300 kilolitres of water per year (1 kilolitre = 1000 litres = 220 gallons). But many households use much more. The following list of typical water usage is taken from *Saving Australia:*

Bath:	50 to 120 litres (11–26 gallons) (half full)
Shower:	40 to 250 litres (9–55 gallons) for the average eight-minute shower
Dishwashing by hand:	18 litres (4 gallons) per wash
Dishwasher:	20 to 90 litres (4–20 gallons) per wash
Clothes washing:	Large automatic machine: 110 to 265 litres (24–60 gallons) per load Twin tub: 40 litres (9 gallons) per load
Handbasin:	5 litres (1.1 gallons)
Tap running while cleaning teeth:	5 litres (1.1 gallons)
Drinking, cooking and household cleaning:	8 litres per person per day
Garden sprinkler:	Up to 1000 litres (220 gallons) per day
Car washing with a hose:	100 to 300 litres (22–66 gallons)
Dripping tap:	Very slight drip 30 litres (6.6 gallons) per day; fine stream 700 litres (154 gallons) per day
Leaking pipe:	300 litres (66 gallons) per day from a 1.5 mm hole.

Did you know . . .

- Twenty per cent of all the toilets in America are leaking right now, and most people don't even know it.
- It takes 13 litres (3 gallons) of water to flush the toilet.
- A dual flush toilet takes 9 litres (2 gallons).
- If a family of four flushes the toilet twenty-four times a day, a dual flush saves 96 litres (21 gallons) a day or 672 litres (148 gallons) a week.

Consider this over a month or year. Impressive isn't it?

NOT A DROP TO SPARE

Courtesy the *Age*, Melbourne

ENVIRONMENTAL POINTS

1   When and how to water plants is important to farmers and gardeners alike.
2   Providing water for animals can be problematical.
3   In many parts of the world, water is in short supply so looking after the water we do have is important.
4   We need to make the best possible use of the water we do have.

WHAT HAPPENS TO WATER?

What happens to water after the rain?

How can you find out?

Plan an experiment to show:

- what happens if you spill a mug of water,

- what happens to the water at different intervals during the day,

- whether this varies with location or time of day,

- what you will try, what equipment you will need and how you will keep a record of your investigation,

- how you will present your findings for best effect.

EXPERIENCE OF THE PROBLEM

A session such as 'Not a Drop to Spare' will provide observations that will arouse curiosity and generate questions that the children will want to explore. The class we worked with noticed and posed the following questions.

'The water on the tiles just sat on top like big drops whereas on the concrete it spread. It dried faster on the concrete. We wondered if this was because the concrete got more sun or because the water spread further.'

'We thought that if we poured the water into smaller amounts rather than one large splodge they might dry quicker.'

'My mum says the wind dries the washing quicker than the sun. We didn't have any wind today so we thought we might try it again on a windy day.'

'There was a crack in the playground and there is still water in it. Everywhere else is dry, we don't know how long it will take to dry out.'

'We drew around the shape of our splodge at twenty-minute intervals. We thought each time it shrank the new shape would be the same only smaller but that's not what happened.'

As the children shared their observations, questions and comments, other children in the class began probing further, asking how and why questions. The children's curiosity and desire to explore further led naturally into an extended project, using this theme. Alternative ways of introducing 'Not a Drop to Spare' could include:

- looking at the water cycle,

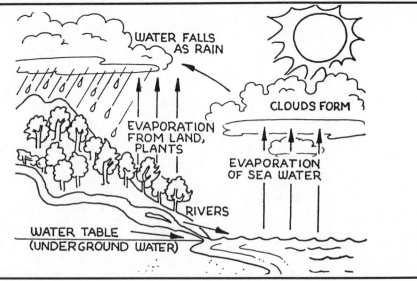

- looking at pictures of floods, rainforests, deserts and discussing issues such as where rain comes from, where it goes, how it gets stored, and why it's essential to farmers and nature,
- investigating songs and rhymes like 'Incy Wincy Spider', 'Rain Rain Go Away', or 'The Rain in Spain', to do with rain,
- researching how different cultures have developed rituals associated with the need for rain, e.g. rain dances, water divining.

MATHEMATICAL ACTIVITY

The earlier experiences and the curiosity that was generated made it possible for us to extend and focus this project by asking the children:

'Why is water so important?'

'Why do some people need to know how much rain falls?'

'How does knowing what happens to the water under different conditions help people like farmers and nurserymen?'

'What information could you collect that might help us understand how we can catch, store and use the water that falls on the ground?'

Using their previous experience and general knowledge, the children developed the following topic web:

We asked the children to form interest groups to explore these topics. Within each group there were different subgroups working on their own ideas which were later compared.

HOW MUCH DOES IT RAIN?

This group put themselves in the role of farmers and decided that weather forecasting, as well as knowing how much rain there had been, would be important to farmers. They investigated weather trends, looking at the sky to see the type, height, density and direction of travel of clouds, trying to establish a pattern that they could use for forecasting. They also checked the daily averages against current weather to see if there were clear links. Their findings were kept on a daily weather chart:

	Clouds	Wind	Average rainfall	Today
Monday 16/4/91	No	Light S. E.	2 1/2 mms	none
Tuesday 17/4/91	high fluffy intermittent	none	2 1/2 mms	none

Meanwhile rain gauges were being made and tested. The children's first thoughts were that any old containers could be lined up and the water in them measured. Trialling showed the inaccuracy of this, so straight-sided containers were sought.

Since there were some days when there was no rain, hence nothing to investigate, we took the children outside to look at local roof shapes and gutters and drainpipes and asked them to set up experiments to explore the effects of different roofs in terms of catchment and strength in the rain.

HOW TO STORE RAIN?

This group went straight into an investigation of storage, using different containers, saucers, dishes, mugs and jugs to hold the same quantity of water. These were placed in the sun and the evaporation was monitored over a period of days. Their first attempt at measuring the evaporation was to measure the depth of the water with a ruler and then compare with earlier readings. This, they slowly realised, did not help them in terms of comparing containers:

'The tall container has gone down 2 millimetres (.08 in). The shallow container has gone down 2 millimetres too so they will dry up at the same time.'

This did not prove to be the case. The measuring technique did not change but the use made of the measurements did. The children found a different way of applying their measurements to straight-sided containers,

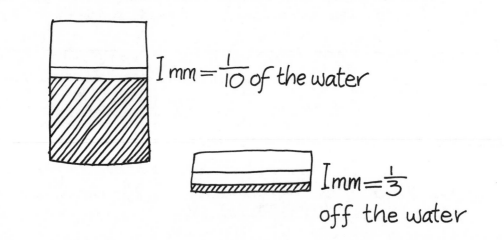

$$I\,mm = \frac{1}{10}\ of\ the\ water$$

$$I\,mm = \frac{1}{3}\ off\ the\ water$$

but even now a problem existed when irregular containers were to be compared,

You can't measure these with a ruler???

Since all containers had been filled with 50 millilitres, pouring into a measuring jug to see what had been lost was tried next.

When the children were coping confidently with this experiment, having resolved how to measure and present their findings, they began to explore evaporation under other conditions — on a cold/damp/windy/cloudy day and in the light and the shade.

HOW TO MAKE IT LAST?

This group began by watering pots of soil. Four pots were filled with dry potting mix. One was given five drops at five-minute intervals; one was given thirty drops at thirty-minutes intervals; one was watered with half of its amount at 9 o'clock and the other half at 3 o'clock; and one was given all its water in one go at 3 o'clock. The moisture was tested at hourly intervals that day and then again the next morning. The children spent a lot of time working on their measuring devices — finger testing, squeeze testing and wrapping a tissue round a paddlepop stick and pushing it into the soil. They decided that the most scientific method was to:

- test with a press of the finger in centre, outside and midpoint,
- test the middle with a tissue on a stick — this can be pasted on the record.

The test was carried out on a coolish, damp day so at 9 o'clock on the second day not a huge amount of difference was noted. The children felt let down but decided to continue the experiment. The second day the experiment continued. By 2 o'clock the 'water once a day' plant pot was quite dry. Noticeable differences were possible after all. So the experiment was continued and findings recorded in a table that showed moisture content over time.

DO ALL PLANTS/ANIMALS HAVE THE SAME NEEDS?

This group decided that the research they needed to do was book and people based. They generated the following list of questions, which were based on their own experiences and general knowledge:

'Do all plants need watering every day?'

'How long can a plant live without water?'

'How much water does a plant need?'

'Do all animals need water every day? We know a camel doesn't.'

'How much water does an animal need?'

'Is it always the same?'

'What plants store water and how?'

They went to the library and phoned and interviewed vets, wildlife park rangers, and nursery employees. One farmer that they were able to speak to told them what a medium-sized dam on his property holds and how much water he needs to irrigate his crops if it doesn't rain.

COMMUNICATING RESULTS

When a reasonable amount of data had been collected by each group, the findings were presented to the rest of the class. At this report-back session, the children were asked to explain:

- what their questions to investigate were,
- what measuring techniques they had tried and how successful they were,
- what they still planned to do.

Comments from the rest of the class helped each group to clarify what else they needed to find out and how the research of one group related to that of another:

'We found that some plants need lots of water and need to be damp all the time, so we could use your information to help us know which watering technique works best.'

'Did you find out if water still evaporates when you put a lid on the container?'

'Did you know some containers like clay pots lose water through the pots?'

It was clear from this session that there was still a lot to find out and a lot of fitting together of each group's work to come to some conclusions about helping to reduce water loss.

REFLECTION

When the children communicated their results, they also reflected on the process of finding ways of measuring and on the relationships between measurement techniques.

'I didn't know you could measure liquid in so many ways. We did areas, how far it spread, centimetres, how deep it was and millilitres which is really how deep it is in the measuring jug. The other group did a finger test to see how wet something is.'

'We noticed that there's two kinds of full, full to the top and full with a raised curve not a flat top.'

'We used time, temperature and area to predict different drying-up times.'

MATHEMATICAL CHECKLIST

As the project unfolded we were surprised by the 'creativeness' of the children in terms of generating ways to measure and ways to record. Just as the children were surprised by how transferable measuring units can be, so we were surprised by their willingness to combine and adapt measurement strategies to suit their needs.

During the course of this topic, you could expect the children to:

Mathematical Topic	Usage
Invent ways of measuring	Water spills and evaporation
Develop ways of recording their observation	As shapes, in measurements, and as charts/graphs
Transfer measurement techniques from their more common usage to new situations	Making linear or area measurements of capacity
Seeing the need for a particular standardised unit of measurement	mls for capacity
Using temperature and timing in combination with other measurements	To conduct experiments
Collect, interpret and represent data	From a variety of written, interview and experimental situations

FACT FILE

Water is used in large quantities in industry. For example, it takes 150,000 litres (33,000 gallons) of water to make 1 tonne (1.1 tons) of steel, and it takes 750,000 litres (165,000 gallons) of water to make 1 tonne of newsprint.

Paris grows about 30 per cent of its vegetables on sewage farms near the city, thus greatly reducing the amount of water required for irrigation.

Australia is the second driest continent in the world.

Lawns and gardens use a vast amount of water. Native plants and grasses once established are not a drain on this resource.

Not all townships grew around a water supply. Los Angeles, for example, obtains its water from rivers 640 kilometres (400 miles) to the north, from the Owen's Valley 400 kilometres (250 miles) to the north-east and from the Colorado River 480 kilometres (300 miles) to the east.

It has been suggested that icebergs be transported from Antarctica to arid regions such as the Arabian Peninsula. This could supply water to 4 to 6 billion people a year or would irrigate 80 million hectares (200 million acres) of land.

Some places now use desalination plants to process saltwater and make it usable.

Careful irrigation avoids water running off the land and seeping into creeks and ultimately out to sea where muddy water silts the beaches.

In the USA in 1980 the total demand for water (excluding water used to generate power) was 1,000 billion litres (240 billion gallons) each day, or 5000 litres (1100 gallons) per person per day. Domestic water use including water used for gardening was about 735 litres (165 gallons) per person.

In the USA 80 per cent of all water used is used for irrigation. Of this water, 65 to 70 per cent simply evaporates and is lost into the air. In hot, semi-arid areas, 2.67 to 6.7 million litres (600,000 to 1.5 million gallons) of water may be used on each acre of crops. Rice paddies require 10 million litres (2.3 million gallons) per acre.

Did you know it takes about...

- 1,600 litres (358 gallons) of water to grow half a kilogram (1.1 lb) of cherries.
- 440 litres (98 gallons) of water to grow half a kilogram of green beans.
- 270 litres (61 gallons) of water to grow half a kilogram of potatoes.

WHEN THE WIND CHANGED

◆

ENVIRONMENTAL POINTS

1 The wind has some good effects — seed dispersal, cooling, breaking up smog — and some not so good effects — rough seas, chill factors, fanning fires, erosion, spreading rubbish — on the environment.
2 Although we can't change the wind we can try to understand its changes and how they relate to our environment.
3 Wind conditions determine what can be planted in some areas. They determine whether it's safe to swim, boat or fish in others. They also contribute to the fire danger in yet other areas. The houses we live in have to be designed to withstand the worst possible local wind conditions — a house only needs to blow down once before it is useless.

BLOWING IN THE WIND

How could you measure the wind? Think about:

- what the wind does when it blows,

- what sort of things it moves,

- how and how far things move in the wind.

What equipment will you need and what procedure will you follow?

EXPERIENCE OF THE PROBLEM

The informal exploration of how to measure the wind will raise new issues that the children will want to explore. The Year 4 children that we worked with noticed that:

> 'There are different types of winds — blustery ones, short sharp winds, long gentle ones.'

and wondered whether they should be measured not as wind speeds but as 'effects'. They also noticed that:

> 'The wind doesn't always come from the same direction. The ones coming from the sea seem to be the strongest ... but we're not sure yet.'

The informal introduction had generated new questions that the children wanted to explore. You may prefer to introduce or extend this introduction by:

- consolidating earlier work on the compass points and making a wind vane,
- using the wave size reports and weather changes to arouse interest in how the wind affects seas and weather patterns,
- going for a walk to observe the effects of the wind on nature, the way trees or plants bend against wind direction or the way the wind blows litter, soil or sand into corners on ridges,
- reading books like *When the Wind Changed, The Hurricane Tree,* and *The Wind Blew,*
- listening to tapes of the noises the wind makes and trying to make wind noises, vocally or with instruments or classroom objects.

We also initiated a discussion about the effects of the wind on people and nature and introduced the concept of erosion.

MATHEMATICAL ACTIVITY

We then asked the children to relate their earlier wind measuring results to the discussion. From this the children's curiosity and need to know generated questions such as:

> 'How far does the wind blow leaves and seeds and spiders' eggs?' (We had just finished reading *Charlotte's Web.*)

> 'Could we be more accurate in measuring the wind?'

> 'How strong does the wind need to be before it causes erosion?'

> 'Boats and kites and things have to be designed to suit different winds — we could invent something that's driven by the wind.'

This enabled us to form four groups to carry the investigation into its next phase.

Group	Topic
How Far?	How far does the wind blow things?
Strength of the Wind	How can we be accurate in measuring the wind?
Erosion	What gets carried away by the wind?
Wind Machine	Can we invent a wind-driven vehicle?

HOW FAR?

This group started by clearing and sweeping the courtyard between two classrooms. They then waited for the next lesson to pass, and collected and marked on a plan whatever was blown into the courtyard. Their idea was to locate where the objects they found had come from and measure the distance travelled. For them, it was quite an experience to find that it was nearly impossible to be sure where a particular object had started, and they realised that they would have to mark both the starting and finishing positions. In the end, they produced an informative chart, showing distances travelled by objects of different weights in a period of ten seconds. They had moved from an informal idea of what it was that they wanted to find out to a well-planned experiment which produced results they could interpret.

'Light things get blown further, but we found a piece of flat paper floats around and changes direction more than a folded or scrunched up bit!'

'In a light wind things seem to roll along the ground, but if it's very windy some extra weight might be needed to stop things blowing up and away.'

STRENGTH OF THE WIND

Although the children had already tried measuring the wind this group now wanted to be more accurate and try to improve on their measuring techniques. In pairs the children developed new ideas, which included:

- noting direction and duration of each gust of wind:

Direction	Time
N	30 sec

- observing and recording things that move, and how, when the wind is blowing.

To them, it was the variation of the wind's strength that aroused the most concern and interest.

EROSION

This group showed how well they had comprehended the notion of erosion when they decided to set out trays of dry sand/sawdust/soil and observe the effects of the wind. They were soon involved in measuring the volume of materials, since they wanted to compare like with like. To their surprise, they discovered that the wind shifted almost all of the sawdust in very little time, but that, unless there was quite a strong wind, the sand and soil remained largely unaltered. The notion that erosion is a slow process was new to them.

WIND MACHINE

When the children began to plan their wind-driven inventions they were able to use information from their research to help them:

> 'It would take a really strong wind to make a model cross the courtyard. Let's just say 10 metres (11 yards) to start with.'

Using collected junk materials, such as cylinders and boxes, and extras like cocktail and kebab sticks, experimentation began. Before trialling their inventions, the children quite spontaneously went outside to gauge the wind so that any last-minute adjustments could be made. The first trialling and in fact most subsequent trials led to new considerations of:

- best materials,
- sail size/material,
- shape,
- angle of wind-catching device.

Although this project is officially over, when last heard from the children were still investigating how best to use the wind as a source of energy.

COMMUNICATING RESULTS

The 'race' was the main focus of communication, as most of the class quickly related their own findings to the practical and interesting problems of the wind machine group. It prompted informal discussion between groups about the best, fastest, most effective vehicles and what to try next. Indeed, the whole class wanted to try their hand at inventing a wind machine.

REFLECTION

The complexities of this project and of course real-life activities like sailing demand that plans for action be changed in the light of new information as it arises, that is reflection has to happen on the job. It becomes second nature to make considerations and to modify plans before a race, as well as while the race is on. Past experience and current experience of the problem, the speed and force of the wind, the properties of the materials used, the design and shape all affect performance.

MATHEMATICAL CHECKLIST

As the children worked on the 'When the Wind Changed' project, they:

- applied measuring techniques that they were confident with to new situations,
- used standard measurements of time, distance and mass to invent new measuring techniques,
- combined information about wind force, materials and construction and shape to design and make their inventions,
- used new information to redesign or modify their inventions,
- integrated spatial aspects of position and shape with measurement and angle,
- explored problems of construction, e.g. joining/fitting shapes together, making moving parts, rigidity and strength.

During the course of this topic you could expect the children to:

Mathematical Topic	Usage
Apply standard measuring techniques, distance, timing, shape, area and mass	To develop new measuring tools and strategies
Use stopwatches, metre sticks, tape measures, trundle wheels	To measure time and distance
Invent measurement scales	To describe the wind's speed/force
Consider characteristics	Of shapes, materials for construction
Consider angle, size and shape	As they design wind-driven vehicles
Construct	Using 'junk' materials

FACT FILE

A man stepped out of his front door, not realising that a tornado had very gently lifted his house 15 metres (16.4 yards) off the ground, and fell to his death.

Planting trees and hedgerows as windbreaks around fields does much to slow soil erosion caused by high winds.

Hurricanes can be dissipated by exploding fireworks near the 'eye' of the storm. The smoke cools the air and causes rain to form, so slowing down the wind and widening the tunnel which forms the eye.

The force of the wind is measured on the Beaufort scale (0–12), but weather forecasters still use terms such as light and heavy to describe wind conditions.

Beaufort Number	Effect on Trees	Kilometres/ hour (mph)	Terms Used by Weather Forecasters
0	Calm, no movement	Less than 1 (1)	Light
1	Very little movement of leaves	1–5 (1–3)	Light
2	Leaves rustle	6–9 (4–7)	Light
3	Leaves and twigs in constant motion	10–19 (8–12)	Gentle
4	Small branches move	20–29 (13–18)	Moderate
5	Small trees begin to sway	30–39 (19–24)	Fresh
6	Large branches sway	40–49 (25–31)	Strong
7	Whole trees in motion	50–59 (32–38)	Strong
8	Twigs break off trees	60–74 (39–46)	Gale
9	Branches break off	75–85 (47–54)	Gale
10	Trees uprooted	86–100 (55–63)	Whole Gale
11	Widespread damage	101–115 (64–72)	Whole Gale
12	Violence and destruction	116–130 (73–82)	Hurricane

This scale was extended to 17 in 1955 to describe a few very unusual occurrences of wind force.

DESIGN A HOUSE

ENVIRONMENTAL POINTS

There are many environmental issues to consider when designing a house:

- how to make best use of the land and its existing features without spoiling any habitats,
- how to make the house a power saver by considering which way to face it to catch the sun or breezes,
- what materials to use so that the house suits the environment and is also energy conscious,
- where roof overhangs are needed for shade and where to place balconies,
- where to install solar heating panels and what size is needed.

ALL I WANT IS A ROOM SOMEWHERE

Design a family room extension for your house.

List all the things you will need to think about.

Draw up a ground floor plan of the room and how it joins onto your house.

Compare your plans with those of other people.

Make any changes that you think would improve your plan.

Why would this be a good plan?

EXPERIENCE OF THE PROBLEM

The children we worked with began to realise that their plans were sometimes too ambitious or that they would not really suit their houses. They also realised that there were shortcomings with their plans and that, in particular, they had no idea about room sizes. Questions had arisen that the children could only answer by doing some research, for example:

'What is a good size for a room?'

'What size rooms do we have at home?'

'Are bigger rooms or smaller rooms better?'

'What is a good shape for a room — square, rectangle or with alcoves and so on?'

'Where in the house should the family room be?'

'What do room/house plans look like?'

The children had begun to see that there were many more issues to be considered than they had first thought, and questions about actually designing a house were being asked.

Other ways of generating interest in and discussion about what's involved in designing a house could include:

* reading books such as *The Hilton Hen House, Tree House Town, The World that Jack Built, A Playhouse for Monster, Amy's Place,*
* walking around the local area to see how houses are designed to cope with the climate and landscape,
* inviting an architect to lend/bring in some house plans or to talk about environmental considerations in the design of a house,
* looking at house and home magazines and the house plans and designs in them,
* researching houses in other times/places,
* brainstorming environmental issues that need to be considered when trying to identify factors relevant to house design and position,
* taking the children to see local building blocks or possible sites and thinking about suitable designs for them,
* developing an awareness of scale drawing by actually marking out a ground plan to full scale on the playground and inviting comments from the children that compare the scale drawing with the full-sized plan.

MATHEMATICAL ACTIVITY

As the children talked their way into wanting to design the ideal home, they generated some very extravagant ideas but slowly came back down to earth, focusing on issues like:

- it can't be too expensive,
- it's got to suit its building block,
- what it should be built of,
- whether to have one or two storeys,
- how to be a power saving house.

The first thing the children wanted to do was to invent a block of land and draw a plan of it. The children had scaled drawings up and drawn shapes to scale, but they hadn't had to scale down before or decide what scale to work to. One group decided that their block would be 320 metres (350 yards) long and 170 metres (186 yards) wide and chose 1m:5cm as their ratio. When they began to convert this it was soon evident that this would not be suitable. Next 1m:1cm was tried, still too large. It was only after moving into millimetres that one child said:

'It would be easier to have more metres to the centimetre.'

and a scale of 5m:1cm was tried. This idea was soon communicated to other groups who were also struggling with scale.

Having decided on their building sites, each group now began to address other issues fundamental to designing a house. These included:

- what size family is it for?
- how many rooms does it need?
- which rooms need to be biggest?
- what is a good size for a kitchen, lounge, bedroom, bathroom?
- which rooms need to be where in the house?
- where should the doors and windows be?
- how many windows and doors does it need?

COMMUNICATING RESULTS

Some groups went straight to designing their houses on graph paper. This led to a few disasters, such as higgledy piggledy looking houses, houses that ended up with lots of corridors or 'dead' spaces, kitchens right next to bathrooms, bathrooms that could only be entered through a bedroom and so on. At this point, we held a report-back session so that the children could talk about what they'd tried and the problems that had arisen. The problems were listed on the board along with some of the children's ideas for solving them. It was now agreed by the whole class that the first stage in design was to 'play' with ideas in rough form first.

MORE MATHEMATICAL ACTIVITY

In their groups, the children now considered different ways of approaching the design problem. These included:

- making cutouts to scale of each room needed and exploring ways of arranging them,
- drawing the house outline to scale and then fitting room cutouts into them,
- drawing roughs and then modifying them before drawing to scale,
- playing around with plans in magazines till they suited the desired need.

At this point, we suggested that each group now consider how their houses would fit their block of land before drawing it to scale. This was necessary because it was obvious that one or two houses would be too wide for the chosen blocks. We felt that at this point the children did not need to face another setback should a beautifully drawn scale plan be found unsuitable for the site.

As the children began to draw their plans to scale, they wanted to be precise in sizes of windows and doors. We asked them to estimate before actually measuring doors and windows in the school. This was worth doing because at first the estimates were a long way off reality. The measuring and the discussion about reasons for differences between estimates and measurements provided the children with the experiences they needed to make these types of estimates.

When the plans were completed to the satisfaction of each group, they of course wanted to make the models of their houses. Construction raised new issues to consider, issues such as:

- how to make sure pieces are cut to exactly the right size and that corners are actually right angles and squares square,
- how to provide support for the roof or ceiling over open plan areas,
- how to join pieces together.

REFLECTIONS

When the houses were complete, we asked the children to comment on what it was like to be an architect and what they had learnt:

'I didn't know so much maths was used in designing and building houses.'

'I never realised how many different ways you could fit eight rooms together.'

'An architect has to see a house in his head, match it to the people who want it and then work it all out. It's hard.'

'I like the way architects have invented signs and symbols like for doors and windows and things.'

'An architect has to plan the house around the living room. It's the biggest best room so it has to be fitted in first.'

When the houses were decorated and on their blocks of land, the children displayed them as a model home village and invited other classes to their exhibition. At the exhibition, they detailed some of the environmental issues to the visitors, explaining why:

- the house faces the way it does,
- the trees had been left on the site,
- one wall had large windows and another small ones,
- why the solar panel was on the roof and positioned where it was.

MATHEMATICAL CHECKLIST

As the children worked on this project, they made some interesting discoveries about:

- logic and the need to deal with the least flexible first,
- using informal maths to explore design possibilities before using formal maths and scale for final drafts,
- the relationship between 2D and 3D representations and shapes,
- decision making where maths determines the decisions.

As the children work on this topic, you could expect them to:

Mathematical Topic	Usage
Use compass bearings	To decide how to face the house
Use linear measurement, area and scale	As they design their block of land and their houses
Use comparison of shape, size and area	As they design rooms and house plans
Experiment with different forms of representation	As they draw or model plans for their houses
See the importance of: • right angles • precise measurements	In design and construction when they begin to construct
Explore shape and area and how the two relate	As they design rooms and house plans
Explore ways of fitting 2D shapes together	To fit 'rooms' into a given area or shape
See the relationship between 2D and 3D shapes	As they construct their models
Explore methods of fitting/joining shapes together	As they make models

FACT FILE

Did you know . . .

- The average house has approximately 1 square metre (1.1 square yards) of 'hole', that is if you add up all the gaps between windows, doors etc., they would total this amount.
- When a block is completely cleared, it costs a lot of money to replant and takes time to establish a new garden. It is better to keep some of the native plants in the first place.
- Each house uses approximately 83,000 megajoules of electricity a year, half for space heating and half for cooking, lighting and working gadgets.
- Timber is the least energy-consuming building material.It takes five hundred times the energy to produce metals and plastics and ten times the energy to produce bricks and mortar as it does to cut and dress the equivalent amount of timber.
- Designing the house to be naturally cool in summer and warm in winter helps to make significant savings in power.

ENVIRONMENTAL TOPIC OUTLINES

- ◆ Baby, It's Cold Outside
- ◆ Eroding Away
- ◆ Shaking All Over
- ◆ Habitats
- ◆ What to Grow
- ◆ Green Shopping

The earlier section presented case studies of environmental topics that we have trialled in the classroom. This section does not outline case studies of activities that we have trialled; rather it provides the skeletons of activities that we have researched and planned ready for classroom use. The activities show how mathematics is used to make sense of the environment and how that maths is monitored on a mathematical checklist.

We include these topic outlines for two purposes: the first to show how we get started planning for a topic and the second to provide a scaffold for teachers wanting to develop their own topic.

The key to working flexibly and taking a process approach is careful planning and preparation. Careful organisation allows us to be flexible, informed, confident and supportive in the classroom rather than directive, judgmental, or worried that the topic is not going to include enough maths or be challenging enough. From our planning we can predict what might happen, what lines of investigation might be taken, what maths might be used and what resources will be required. These are our advance organisers. They also enable us to check before a project gets underway that it will provide the necessary opportunities for the children to:

- acquire new skills,

- practise and transfer the skills they already have,

- extend existing knowledge or concepts,

- use mathematics creatively,

- use maths to get things done or make things happen.

Worries about 'covering the syllabus' need not then arise. We have shown on the mathematical checklists the kind of maths that might be used. This acts not only as a check but also as a means of anticipating areas where some direct teaching may be needed.

By including this planning here we hope to provide activities that are not 'prescribed', activities that you can adapt to the needs of the children in your class and in a way that suits your teaching style. We also hope to provide a model for those who want to plan their own environmental topics.

As in the case studies, we would urge you to:

- encourage children to generate their own questions and areas to investigate,

- encourage creativity in approach and risk taking,

- use report backs at all stages of the topics so that the children can share ideas, extend ideas and benefit from other experiences and strategies,

- ensure that the children's recommendations and results are shared, published or used to bring about change,

- provide time for the children to reflect on what they have done, what worked and didn't work and why, what they have learned, what they will do next or do differently next time.

Baby, It's Cold Outside

ENVIRONMENTAL POINTS

1 Temperature can affect plant growth; if it is too hot plants burn and wither, if it is too cold or frosty they die. Different plants have different needs in terms of light and shade, temperature and moisture requirements.

2 Our bodies react to temperature — we don't like to be too hot or too cold.

3 Buildings are designed to keep heat in or out *or* cold in or out depending on the part of the world we live in.

4 The earth is said to be warming and this could cause tremendous changes to the world as we know it.

5 Temperature determines how long food will keep and how effectively people will work.

SOME LIKE IT HOT

What's a good temperature for a classroom?

How often/when is your classroom the right temperature?

What determines the temperature in your classroom?

Is it different at different times of day? Why?

How/where/when will you measure the temperature?

What equipment will you need?

What is there about your classroom that helps to keep it cool/warm?

What changes would you suggest to your classroom's design or position to make it a better work environment?

DEVELOPING THE TOPIC

As the children explore the temperature inside the classroom, encourage them to keep records of their findings as they work, for example:

	Inside		Outside		
	Near window	Mid-room	Sun	Shade	Wind
9.00 am	25°	24°	22°	21°	none
10.00 am	27°	26° fan on	24°	23°	none
11.00 am	27°	26° fan on	26°	26°	none
12.00 am	28°	26° fan on	27°	27°	light
1.00 pm	28°	26° fan on	27°	26°	light
2.00 pm	27°	25° fan on	28°	26°	none
3.00 pm	25°	25° fan on	27°	25°	none

Note: This class recorded their temperatures in ^{o} Centigrade.

and to use this information to draw conclusions about how efficiently their classroom has been designed:

'Our room is pretty well designed because most of the time it's cooler inside than out. We'd need to see what it's like on a cold day though. We think blinds would help in the morning when the sun glares in.'

They may raise issues or questions like:

'Why is it cooler/warmer in here than it is outside?'

'How do sunshades help?'

'If the classroom was turned round 90 degrees we wouldn't get the sun in the morning.'

'The temperature is lower/higher outside than it is inside at some times during the day.'

RELATED QUESTIONS

As the children explore the temperature around their room, other related questions that they want to explore may arise, for example:

'We know 28 degrees is too hot for work. But would it be good for swimming?'

'When is it too cold/too hot for comfort?'

'Do some people cope with heat/cold better than others?'

'What makes you cool down/warm up?'

'When you get hot/cold, does your body temperature go up/down too?'

Body sense can provide openings too:

'Put your finger in a jar of warm water then into iced water. What do you notice?'

'How much difference in temperature does there need to be before you can notice it? Move around the room to find a hot spot/warm spot/cool spot. How much difference is there before you notice the changes?'

'What is a good temperature for porridge/coffee/Coke?'

'What sort of temperature variations do you notice between night and day, a.m. and p.m.?'

'Why do we measure the temperature of a baby's bath water with our elbow?'

'Which is the most temperature-sensitive part of our body?'

Another alternative would be to read books or stories where temperature is part of the plot or theme, e.g. *Mr Snowman* or *The Little Match Girl*.

MATHEMATICAL CHECKLIST

During the course of this topic you could expect the children to:

Mathematical Topic	Usage
Estimate temperatures	To notice changes or predict current temperatures
Invent the scale for a thermometer and see the need for a standardised scale	Using an unmarked thermometer to find hottest/coolest spots (fridge and sunny spot)
Learn how to take temperatures	As they realise that it takes time for the mercury to rise or fall
Use a standard thermometer	To measure and record different temperatures
Record times and temperatures as charts, tables, grids	As they keep a record of their results
Compare temperatures at different times, different places and under different conditions	To make sense of their environment
Use time, position and temperature	In written reports and as parts of maps or plans

FACT FILE

Did you know . . .

- The inside of a cucumber is 11°C (20°F) cooler than the outside temperature.
- In Spearfish North Dakota on 22 January 1943, the temperature rose from -4°F at 7.30 a.m. to 49°F at 7.35 a.m., in just five minutes.
- Lowering the temperature setting on your water heater could save 10 per cent of the electricity bill.
- If everyone in the United States turned down their heat by 6°F, half a million barrels of oil could be saved every day.
- Insulation prevents heat from escaping; good insulation can save 30 per cent of the heat loss.
- If your house has blank walls facing east and west much of the sun's heat is not collected.
- In Australia, overhangs on the north side of the house let the warmth in in winter but keep the sun out in summer.
- If your house is built to catch the breezes to the north and south it will be appreciably cooler than a home that is not built in this way.

ERODING AWAY

ENVIRONMENTAL POINTS

1   There are many causes of erosion. Some are natural processes that
   wear away the land areas of the world, such as the effects of the weather,
   wind and rain and glaciers. Other effects are caused by humans and
   can be avoided, e.g. land clearing and overfarming.
2   Careful land clearing, reforestation and planting can counter some
   of these effects.
3   Staying on paths etc. when visiting beauty spots will help to preserve
   natural habitats.
4   Replacing naturally what we take from the earth for crops and animals
   will help counter erosion.

WASTING AWAY

What can you find out about the way the wind or rain causes erosion?

Set up an experiment using the items above.

How will you set up the experiment so that the starting conditions are all the same?

How will you measure the changes each time?

How will you record your findings?

DEVELOPING THE TOPIC

As the children explore erosion, they may raise other related questions or issues that they will want to explore further, for example:

'Erosion happens when people walk on grass and the path turns to mud or dirt. Why is that?'

'Some building materials like sandstone rot away. Which are the toughest materials?'

'When it rains hard the water runs off the surface making little rivers. How hard does it have to rain before this happens?'

Other ways of introducing erosion could include:

• going for a walk or visit around your area to see some recent land clearing for quarrying, building, road works, farming or whatever,
• reading books with related themes, e.g. *Norah's Ark, Conservation, Goanna, The Machine at the Heart of the World,* or *Mr Shanahan's Secret,*
• looking at pictures of places created by erosion — canyons, rivers, or avalanches,
• walking around the school grounds looking for changes due to the weather or wear and tear,
• initiating a discussion about erosion by inviting the children to talk about the erosion they've experienced.

The children may have comments and suggestions such as:

'We lost our gravel in the rain, it was all on the road.'

'There's less grass growing where people walk a lot.'

'The trees on the hill are bent away from the wind.'

'The steps are sort of worn out in the middle where people walk.'

Use the children's suggestions to help them ask questions that they can then investigate. If necessary ask questions yourself to help the children get going, for example:

'How could you find out what effect rain has on different types of areas/materials?'

'What does the wind do to plants and the soil?'

'What material makes the best paths/steps/garden walls? Why?'

'What happens when you pour water onto a sand castle?'

'What happens to the countryside when lots of people walk all over it?'

'Is there anywhere around the school grounds that is being changed by use or the weather?'

'How could you collect information about what causes changes?'

'How could you measure the changes?'

MATHEMATICAL CHECKLIST

During the course of this topic you could expect the children to:

Mathematical Topic	Usage
Develop measuring techniques (e.g. using capacity, litres, mass, grams, kilograms, depth in a bucket or on the tray arms)	To conduct their experiments
See the need for more precise measurements	As they begin to want more accuracy in their results
Compare results	What is changed, moved under different conditions
Develop ways of recording and presenting information	For reports
Integrate timing and speed of flow or wind mass, capacity and distance	As they refine their experiments (is 1 litre poured slowly, say in ten minutes, more damaging than 1 litre poured in thirty seconds?)

FACT FILE

The Australian wheat crop of 1986-87 was 16.1 million tonnes (1 tonne = 1.1 tons) and it removed from the soil:

  322,000 tonnes of nitrogen
   48,000 tonnes of phosphorus
   64,000 tonnes of potassium
   32,000 tonnes of sulphur

Half of Australia's agricultural land is affected by erosion.

Deforestation in some parts of the world leaves poor soil, which is only good for growing crops for two to three seasons. After this, more forest has to be cleared.

A forest the size of a football pitch disappears every second.

There are 1.5 billion hectares (3.7 billion acres) in the world, and 2 million hectares (4.9 million acres) are completely ruined by erosion each year.

When planning a garden the use of native plants helps to retain the soil and to enrich it — that's why they were native to that area in the first place.

Did you know . . .

- Desertification, the turning of plant-bearing land into desert, has become a major world problem.
- One-third of the world's land is already desert, or affected by desertification.
- Every year the Sahara desert spreads to include an area about the size of Czechoslovakia (120,000 square kilometres or 46,000 square miles).
- The Soviet Union and Australia are the two countries with the most land at risk from desertification.

SHAKING ALL OVER

ENVIRONMENTAL POINTS

1 Although most areas are not prone to earthquakes, buildings have to be designed to withstand all kinds of 'acts of God'.

2 By exploring building materials and constructions, we can develop an understanding of the many different ways our homes protect us from the elements.

3 Mining, blasting and underground nuclear testing can cause earthquakes. Nuclear testing can generate earth tremors that can be felt for 150 kilometres (93 miles) or more.

CREATE AN EARTHQUAKE

How could you measure an earthquake?

How could you set up an earthquake experiment?

What equipment would you need?

What do you think happens in an earthquake?

What would you do to simulate an earthquake?

How could you measure an earthquake?

DEVELOPING THE TOPIC

As the children create their own mini-earthquakes, they may raise other related questions that they will want to explore. These could include:

'What sort of buildings are safest in an earthquake?'

'Is there a better way of measuring shock waves?'

'Does it make any difference what building materials are used?'

'I heard that sand makes a good foundation for buildings where earthquakes happen. Is that true?'

The starter activity will have given the children an idea of what earthquakes are and what they are like. The focus can now broaden to include the way buildings and structures are constructed to withstand as far as possible 'acts of God'. Ways of introducing the topic could include:

- going on a walk around the school grounds or local area to explore building materials, structures, designs and orientation or position used in local homes,
- initiating discussions about how a house's design helps to protect its inhabitants from the elements, e.g. why many houses don't have flat roofs,
- reading stories or books that have natural disasters as their theme, e.g. *Norah's Ark, Changes, Changes, The Three Pigs* (an unnatural disaster),

If necessary you might want to suggest further explorations to stimulate further questions, for example the children could try:

- building walls with blocks, milk cartons, soap powder boxes and other materials and frames,
- creating an earthquake to see which types of construction withstand an earthquake best,
- exploring building surfaces, e.g. sand, gravel, blocks,
- making some roofs and seeing which are strongest or which deal best with a heavy rain,
- exploring which building materials will keep the water out best — bricks, concrete blocks or wood — marking the level of water before the object is immersed and again after the object has been immersed for some time and then removed is one technique for this,
- seeing what they can discover by building two/three/four storey buildings with assorted materials.

MATHEMATICAL CHECKLIST

As the children explore how our homes are built to withstand earthquakes and the elements you could expect them to:

Mathematical Topic	Usage
Invent ways of creating and measuring earthquakes	As scales, or by measuring movement
Time the duration and force of earthquakes	To create standard experiments
Invent ways of recording the data collected	As graphs, tables, plans, lists of measurements
Explore patterns in construction	How bricks are fitted together for maximum strength
Explore the effects of roof pitch	To see which are strongest, most effective in withstanding rain
Explore methods of construction (fitting, joining)	As they build with a variety of materials
Compare materials in terms of 2D, 3D shape/size/ material/strength	As they construct
Use water displacement to measure absorption	Of different building materials
Use mathematics to explain and persuade	As they report on best materials and designs for building in an earthquake zone

FACT FILE

The largest earthquakes are sometimes felt more than 1,600 kilometres (1,000 miles) from the source.

Shock waves can be felt on seismographs on the other side of the world.

Surface waves travel at 3 to 4.4 kilometres (1.9–2.7 miles) a second.

Earthquakes can be measured on a seismograph using the Richter scale, summarised as follows:

Magnitude	Effect
2	Smallest shocks normally felt
4.5	Smallest shocks causing slight damage
6	Moderately destructive
8.5	Largest shock known

Earthquakes can also be measured in terms of damage done to manmade constructions. This is a US-modified form of the Mercalli scale (1931):

Magnitude	Effect
1	Felt by a few under especially favourable conditions
2	Felt quite noticeably indoors. Vibrations like those of a truck passing by
3	Dishes, windows, doors disturbed, walls make cracking sounds
4	Felt by nearly everyone: many awakened. Unstable objects may be overturned
5	Felt by all. Damage slight
6	Fall of chimneys, monuments, walls. Changes in well water. Considerable damage in ordinary substantial buildings
7	Most masonry and frame structures destroyed, landslides
8	Damage total. Shock waves are seen on the ground's surface

MAJOR DISASTERS

Lisbon earthquake 1755	20 per cent of population killed
	50 per cent of city destroyed
	3,890,000,000 square kilometres (1.5 billion square miles) felt the waves (most of Europe)
San Francisco 1906	500 people killed
	Damage estimated at $20,000,000
	Fire damage at $500,000,000
Alaska 1964	One of largest ever recorded, generated huge sea waves that travelled as far as Hawaii and Hokkaido, Japan
	Few killed
	Damages estimated at $500,000,000
Peru 1970	50,000 people killed
	200,000 people left homeless
Nicaragua 1972	More than 12,000 people killed
	75 per cent of buildings destroyed
	300,000 people evacuated
Tangshau (China) 1976	Estimated 650,000 people killed
	Peking, 160 kilometres (100 miles) north-west of Tangshau, also sustained heavy damage to property

HABITATS

◆

ENVIRONMENTAL POINTS

1　Every time we chop down a tree, replace a native tree with a non-native or clear land for a house, we are changing the balance in the local ecosystem.
2　Only by observation and investigation can we really develop an understanding of the habitats that even our backyard or school ground harbours.
3　From this awareness we can begin to make plans to try to preserve some of these habitats and ecosystems.
4　Species of plants and animals are disappearing from our earth every day.

WHAT'S IN A TREE?

How could you find out what lives in/on a tree?

What will you need to do?

How will you make sure you don't miss anything?

What equipment will you need?

Note: If yours is an inner city school with no trees, don't despair. Find out what lives in/on a wall or crack or cranny.

DEVELOPING THE TOPIC

The children will be surprised by the extent of the wildlife to be found in a shrub, tree or cranny and may generate other aspects or questions that they want to explore. These could include:

'Are there more creatures or plants in the shade or in the sun?'

'Do some types of plants have more creatures in them than others?'

'What about visitors/fires/bees/birds/snakes/possums and so on?'

'If we put a new plant outside, how long would it take before it has a big creature population?'

This starter activity will engage the children and help develop an awareness of how a plant can be a mini ecosystem. An investigation of different areas and the range of wildlife supported by them could follow.

RELATED TOPICS

Other ways of broadening this focus could include:

* reading books or stories such as *Father Sky and Mother Earth, Adam and Paradise Island, The Last Dodo, Antarctica, Rainforest,*
* visiting a zoo or wildlife sanctuary to see how habitats are created,
* walking around your local area to see what provisions your local authority has made for 'nature strips' or wildlife sanctuaries,
* scratching around tree roots and seeing what lives in the soil,
* setting up a mini-pond and seeing how quickly it becomes inhabited.

MATHEMATICAL CHECKLIST

As the children explore different types of habitats you could expect them to:

Mathematical Topic	Usage
Keep a tally	Of the species they find
Find ways of recording their information	As tables, lists, graphs, plans, pie charts
Compare and average	The types of creatures found in different habitats
Use repeated addition, multiplication, estimation and rounding	As they try to get a feel for how many inhabitants a tree or an area might be housing
Take samples	To predict longer term visitors to a tree or area
Use ratio or averages	To expand results from a small area to a larger one

FACT FILE

Did you know . . .

- Many wetlands, rivers, and underground resources are polluted so native flora and fauna are disappearing.
- In Brazil, there are only four hundred marmosets left because of the rate at which forest is being destroyed.
- Following the 'country code' helps to preserve habitats.
- Antarctica is 12,000,000 square kilometres (4,600,000 square miles) of unpolluted, unexploited wildlife sanctuary. This is because people have not been there to spoil it. However it is rich in minerals which people want to mine.
- Overfishing upsets the balance in our oceans. Krill, which is the key to the oceans' ecosystem, is in danger of being overfished.
- In 1900, there were an estimated 400,000 tigers. By 1930, there were only 40,000 tigers. By 1970, this had fallen to 4,000 tigers. In 1972, tigers were protected and Project Tiger was launched in India to try to provide safe habitats for tigers.
- In England, small fields have been made into large ones by removing hedgerows that are very old. A quarter of all hedgerows and their habitats have now been destroyed. That is 750,000 kilometres (466,000 miles) of hedgerows or about 4,000 kilometres (2,500 miles) a year.

WHAT TO GROW

◆

Courtesy the *Age*, Melbourne

ENVIRONMENTAL POINTS

1 Deforestation for timber, farming and building has endangered many
   species whose habitats have been destroyed in the process. It is
   important therefore that we begin planting trees and shrubs that will
   provide food and homes for our indigenous species.
2 Most native plants will grow well where they are found naturally. The
   soil balance will be appropriate for them and so too will the climate.
   This means that the need for watering, fertilising and pesticides will
   be reduced.
3 'Companion' planting can reduce the incidence of pests and diseases
   in plants too.

ALL'S WELL THAT GROWS WELL

What grows well around here?

How can you find out?

Make a plan showing:

- what information you need to collect,

- how you will collect it,

- how you will record your findings.

DEVELOPING THE TOPIC

As the children investigate which plants grow well in their area, they may begin to ask questions that they want to explore further. Questions such as:

'Are the most popular plants actually the easiest to grow?'

'Which plants are natives?'

'Why are some plants of the same type doing less well/better than others.'

'What is the tallest/bushiest?'

'Why have some plants finished flowering or have fewer flowers than others of the same kind?'

This starter activity will have given the children an awareness of why some plants are easier or more popular in one area than others and will lead into broader topics such as 'Planting a nature strip' or 'What to grow to improve our classroom'.

RELATED TOPICS

Alternative ways of introducing this topic could include:

- reading books or stories such as *The Lorax, Johnny Appleseed, Isn't It a Beautiful Meadow?, Kenju's Forest, The Giving Tree, Apple Pigs, A Forest Is Reborn,*
- visiting a different type of area (exposed hillside, seaside, delta, rocky, clay based) to see what grows well there,
- visiting a nursery to see what species are being sold and to interview the gardener,
- visiting a local botanic garden or park,
- researching native plant species in books,
- looking at weeds and how and where they grow,
- setting up experiments to explore how plants spread (seed type and disposal, root systems, propagation),
- observing and recording 'visitors' to the plants to see what insect or bird species they attract.

MATHEMATICAL CHECKLIST

As the children explore how plants grow, how and why they do well in a particular area, and which wildlife they attract they might be involved in:

Mathematical Topic	Usage
Collecting data about plant location	To find out which plants like shade, light, heavy or light soil, rocks, wind, damp or dry areas
Recording data on plans or maps	To show how many plants of each type and their most common growing conditions
Inventing measuring techniques to establish how 'healthy' plants are in particular conditions	These could include height, density of leaves, flowers, girth, number of main and subsidiary branches
Averaging	To find the most common size of plants
Observing, timing and recording data about	Number of visitors to a plant, duration of stay, most common time of visits
	Seed dispersal, types of seeds and how far they travel
Measuring and comparing results of their own	Growth experiments and means of propagation

FACT FILE

Did you know . . .

- Since the 1960s 1.5 per cent of the world's tropical rainforests have been cut down each year. That's about 8 million hectares (20 million acres) or 8 million trees a year.
- In Australia two-thirds of the natural tree cover has been destroyed.
- In 1950, 16 per cent of the earth was covered in tropical rainforest, now there is only 10 per cent tropical rainforest and in forty years time, there will be NONE! It would take four hundred years to grow it all back.
- Weeds should be removed because they use a lot of the water and nutrients needed by the other plants.
- Plants need to be 'toughened up'. They get so used to being watered that they don't look for water themselves by sending their roots down deeper.
- Lawns use more water than plants, so large lawns are not always practical.

GREEN SHOPPING

◆

ENVIRONMENTAL POINTS

1 When we go shopping we take home too much packaging.
2 Packaging wastes natural resources or pollutes the atmosphere in its production because of the use of fossil fuels in their production and because of the waste byproducts that flush into our waterways.
3 When we shop we are often actually paying for the packaging not the product, choosing a brand or label rather than the 'best buy'.
4 Most packaging ends up as rubbish.
5 Many of the domestic products that we use are not biodegradable; they enter the sewage and then filter into the land or out to sea.

A TRUE STORY

In the plastic carrier bag that the chemist gave us there was a paper bag. And in that paper bag there was a plastic wrapping. And in that plastic wrapping there was a cardboard box. And in that cardboard box there was a polythene liner. And in that polythene liner there were one hundred individually wrapped bandaids!

That's six layers of packaging between us and a bandaid. Is that necessary?

How much packaging are we paying for?

Choose a common item from the supermarket. Find out how much wrapping is on it.

How much of it is essential/not essential?

Are you buying more packaging than product?

What else is packaging for?

DEVELOPING THE TOPIC

As the children investigate just how much packaging we buy every time we go shopping, they will become aware that:

- much of it is unnecessary,
- the packaging is often deceptive with large packets containing a little product and a lot of air,
- the packaging is often more attractive than the product within.

Questions and comments arising from these realisations could include the following categories:

- best buy or best value for money,
- best way of packaging without wasting materials,
- alternatives to packaging,
- how much packaging could manufacturing actually save if they made packets an appropriate size and didn't have too many layers.

MATHEMATICAL CHECKLIST

As children investigate packaging they might be involved in:

Mathematical Topic	Usage
Collecting data about mass and capacity (grams, kilograms, litres etc.)	To find out how much each container actually holds
Comparing amount of volume filled with volume available	As they explore how deceptive some packaging is
Weighing or quantifying packaging to classify type and quantity of material used	To collect information about how packaging is necessary or superfluous
Exploring properties of 2D and 3D shapes and their relationships	As they take packaging apart and/or design more practical packaging
Combining money and quantity or measurement	As they explore value for money

FACT FILE

Did you know . . .

- For every forty-five cans recycled, thirty-eight new ones are made.
- Aluminium cans are thinner now than they were ten years ago. This means that an extra seven thousand cans can now be made from each tonne of aluminium.
- In the USA 1,000,000 tonnes of aluminium is used to make cans every year.
- Glass is made from 50 per cent sand.
- Paper with plastic or wax on it cannot be recycled.
- One tree can provide the pulp to make four hundred and twelve newspapers.
- The average American uses seven trees a year, in paper, wood and other products made from trees.
- Fourteen million copies of the Japanese newspaper, *Asahi Shimbun*, are sold every day. That's about 34,000 trees!
- Americans use 45 million tonnes (50 million tons) of paper every year, or about 270 kilograms (580 lb) per person.
- If everyone in the United States recycled their Sunday papers, we would save 500,000 trees every week.
- Many foods have artificial (chemical) food additives.
- Pressurised cans may contain CFCs (chlorofluorocarbons) and even those that don't have hydrocarbons that add to the greenhouse effect.
- Overfishing, such as catching tuna in 50 kilometre (30 mile) long trawling nets, is endangering tuna and killing dolphins, seabirds and anything else that gets caught in the nets.
- If vegetables were labelled we would have some idea of how 'organic' they are. Suggested categories are:
  Grade 1:   No synthetic fertilisers, pesticides, growth regulators,
  (green)   antibiotics or hormone stimulants
  Grade 2:   As for Grade 1, except some synthetic products may have
  (orange)   been used on the soil in the recent past
  Grade 3:   Small amounts of synthetic products may have been used
  (red)   with care
- You can use bicarbonate of soda as a cleaning agent. It's good for teeth cleaning too.
- Information on labels could advise on how to use or dispose of certain products.
- Japan is being called the Throwaway Society because no one repairs anything anymore.
- Buying the best pays off. It usually lasts longest and is cheaper and kinder to the environment in the long run.

THINK ABOUT IT,
ACT UPON IT

The topics suggested in this book will have given the children insights into the way that maths can help to explain how our universe is being spoilt and also how it can help us work out what to do about it. The areas covered will provide the background information needed to explain, report on, persuade and clarify issues like those that follow. Get ready for some heated discussions, some brainstorming and some creativity as the children tackle topics like these:

- Should households be charged for the water they use?
- Should households be charged for the rubbish they have taken away?
- Would paying for recyclable rubbish as it is collected be worthwhile?
- Petrol should be rationed — yes/no?
- We should leave all packaging at the supermarket.
- Houses should be made of rubber.
- Electricity should be graded in price: low tariff for low usage, high tariff for high usage.
- What is a good shopper?
- Careful shopping can help save the earth — true/false?
- Find one hundred uses for the empty ice-cream container/cardboard box/polythene bag ...

The earlier topics will also provide the background experience and mathematical knowledge to ensure that a 'Design a ...' project becomes a practical possibility. For example, you could try the following 'Design a . . .' or 'Invent a . . .' topics:

- Design a play area that is environmentally friendly.
- Invent a way to turn the most barren, arid or ugly part of your school yard into a lovely lush area.
- Design and create a nature trail or nature maths trail around your school.
- Invent an environmentally friendly shopping kit.
- Design a herbarium or nature strip.
- Invent a compact portable compost maker.

Maths in Context
Deidre Edwards

Deidre Edwards, an experienced teacher of primary school mathematics, shows how to integrate mathematics with the wider curriculum areas by using a central theme. As a result of this approach, she demonstrates that there is an increase in children's motivation, individual differences are catered for, children's confidence in their mathematical ability grows and mathematics is seen as part of 'real life'.

Maths in Context provides guidelines on how to handle group work, on classroom organisation and on planning and implementing assessment strategies.

A large section of the book presents ideas for activities based on the following themes: Dragons, Our Environment, The Zoo, Party Time, Traffic, Christmas, Show and Tell, the Faraway Tree.
Illustrated 152 pp

For further information about these and other exciting books for teachers contact:

Eleanor Curtain Publishing
2 Hazeldon Place
South Yarra Vic. Australia 3141
Telephone (03) 826 8451 Fax (03) 827 3115

Ashton Scholastic
165 Marua Road
Auckland 6
New Zealand
Telephone (09) 579 6089 Fax (09) 579 3860

Heinemann Educational Books Inc
361 Hanover Street
Portland
NH 03801–3959
Telephone (603) 431 7894 Fax (603) 431 7840

RELATED BOOKS

Material for the Fact Files has been taken from the following publications, and from *Collier's Encyclopedia*:

Javna, John & The EarthWorks Group. *50 Simple Things Kids Can Do to Save the Earth*. Andrews & McMeel, 1990.

Serventy, Vincent. *Saving Australia*. Child & Associates, 1988.

——. *Your Easy Guide to Green Living*. Child & Associates, 1990.

The following books, which have been referred to in the Case Studies and Environmental Topic Outlines, are suitable for use in the primary school.

Note: Where available, US editions of the books have been provided. Where no US edition is available, British and Australian details have been provided.

Allen, Pamela. *Bertie and the Bear*. Putnam, 1984.

——. *Mr Archimedes' Bath*. Nelson, 1983.

Andersen, Hans Christian. *The Little Match Girl*. Word Books, 1986.

Ayres, Pam. *When Dad Fills in the Garden Pond*. Knopf, 1988.

——. *When Dad Cuts down the Chestnut Tree*. Knopf, 1988.

Baker, Jeannie. *Where the Forest Meets the Sea*. Greenwillow, 1988.

——. *Window*. Greenwillow, 1991.

Banner, Angela. *Ant and Bee Go Shopping*. Trafalgar Square/David & Charles, Inc., 1988.

Briggs, Raymond. *Mr Snowman*. Random, 1986.

Brown, Ruth. *The World that Jack Built*. Andersen, 1990.

Burningham, John. *Hey! Get off our Train*. Crown, 1990.

Cartwright, Ann & Cartwright, R. *In Search of the Last Dodo*. Little, Brown, 1989.

——. *Norah's Ark*. Puffin, 1985.

Cowcher, Helen. *Antarctica*. Farrar, Straus & Giroux, 1990.

——. *Rain Forest* (includes audiocassette). Soundprints, 1989.

Dunkle, Margaret. *Conservation*. Puffin, 1989.

Facklam, Margery. *And Then There Was One: The Mysteries of Extinction*. Sierra, 1990.

Flanagan, Joan & Seal, B. *Mr Shanahan's Secret*. Gareth Stevens Inc., 1988.

Foreman, Michael. *Dinosaurs and All that Rubbish*. Puffin, 1974.

Gates, Richard. *Conservation*. Children's Press, 1982.

Hargreaves, Roger. *Mr Noisy*. Price Stern, 1980.

Harranth, W. & Opgenoorth, W. *Isn't It a Beautiful Meadow?* Translated by Jeffrey Tabberner & Ron Heapy. Oxford, 1987.

Hinchcliffe, Jo. *The Hilton Hen House*. Ashton Scholastic, 1982.

Hutchins, Pat. *Changes, Changes*. Macmillan, 1987.
———. *The Wind Blew*. Penguin, 1986.
Keeping, Charles. *Adam and Paradise Island*. Oxford, 1989.
Kellogg, S. *Johnny Appleseed*. Morrow, 1988.
McCully, A. *Tree House Town*. Little, Brown, 1974.
Miles, Betty. *Save the Earth! An Ecology Handbook for Kids*. Knopf, 1974.
Morimoto, Junko. *Kenju's Forest*. Golden Press, 1989.
Mueller, Virginia. *A Playhouse for Monster*. Penguin, 1988.
Newton, J. R. *A Forest Is Reborn*. Harper & Row, 1982.
Orbach, Ruth. *Apple Pigs*. Putnam, 1981.
Park, Ruth & Niland, Kilmeny. *When the Wind Changed*. Nelson, 1984.
Punt, I. & Holdcroft, T. *The Bop*. Ashton Scholastic, 1989.
Purves, L. & Lamont, P. *The Hurricane Tree*. Bodley Head, 1988.
Roennfeldt, Robert. *Tiddalock Frog Caused a Flood*. Penguin, 1983.
Ross, Tony. *The Three Pigs*. Pantheon, 1983.
Seuss, Dr. *The Lorax*. Random, 1971.
Silverstein, Shel. *The Giving Tree*. Harper & Row, 1964.
Stafford, Marianne. *Amy's Place*. Nelson, 1983.
Wagner, Jenny. *Goanna*. Penguin, 1989.
———. & Fisher, Jeff. *The Machine at the Heart of the World*. Puffin, 1983.
Wakefield, S.A. & Digby, Desmond. *Bottersnikes and Gumbles*. Golden Press, 1988.
Walker, Kath. *Father Sky and Mother Earth*. Jacaranda, 1981.
Wildsmith, Brian. *Professor Noah's Spaceship*. Oxford, 1980.

More ideas to inspire your teaching of mathematics:

Raps & Rhymes in Maths
Compiled by Ann and Johnny Baker

A collection of traditional and modern rhymes, riddles and stories with mathematical themes, *Raps, Rhymes in Maths* can be used to provide a welcome break from more formal rhymes and stories provide openings for mathematical investigations and, most importantly, provide a source of enjoyment.
Illustrated 90 pp

Mathematics in Process
Ann and Johnny Baker

Mathematics in Process extends the confidence that teachers now feel in the language arts to the mathematics lesson. The purposes and conditions of natural learning, now common in the language classroom, are applied to learning and doing mathematics.
    This very comprehensive book is divided into three parts:
• Part One looks at the child's experience and has sections setting out how children get involved, how young mathematicians work, how children communicate and learn from reflection.
• Part Two sets out classroom approaches: identifying purposes for using mathematics, the conditions for learning mathematics, shared experiences for learning mathematics and assessment.
• Part Three relates ideas on devising a curriculum, how to set up the classroom and features a complete section on activities to try with your class.
Illustrated 176 pp

Maths in the Mind
Ann and Johnny Baker

Maths in the Mind focuses on the development of mental skills and strategies within the context of broader activities and emphasises the processes involved in mathematical thinking. Children are encouraged to explore, to delve into what they do know rather than feel helpless because of what they don't know. Twenty fully developed activities are set out in the book.
    Maths in the Mind complements *Mathematics in Process*, also written by Ann and Johnny Baker.
Illustrated 120 pp

Maths in Context
Deidre Edwards

Deidre Edwards, an experienced teacher of primary school mathematics, shows how to integrate mathematics with the wider curiculum areas by using a central theme. As a result of this approach, she demonstrates that there is an increase in children's motivation, individual differences are catered for, children's confidence in their mathematical ability grows and mathematics is seen as part of 'real life'.

Maths in Context provides guidelines on how to handle group work, on classroom organisation and on planning and implementing assessment strategies.

A large section of the book presents ideas for activities based on the following themes: Dragons, Our Environment, The Zoo, Party Time, Traffic, Christmas, Show and Tell, The Faraway Tree.
Illustrated 152 pp

For further information about these and other exciting books for teachers contact:

Eleanor Curtain Publishing
2 Hazeldon Place
South Yarra Vic. Australia 3141
Telephone (03) 826 8451 Fax (03) 827 3115

Ashton Scholastic
165 Marua Road
Auckland 6
New Zealand
Telephone (09) 579 6089 Fax (09) 579 3860

Heinemann Educational Books Inc
361 Hanover Street
Portland
NH 03801-3959
Telephone (603) 431 7894 Fax (603) 431 7840